MW01225479

Tell Me Why

WHY?

Flowers Bloom

Nancy Robinson Masters

Published in the United States of America by Cherry Lake Publishing
Ann Arbor, Michigan
www.cherrylakepublishing.com

Content Adviser: Elizabeth A. Glynn, Youth Education Coordinator, Matthaei Botanical Garden and Nichols Arboretum, University of Michigan
Reading Adviser: Marla Conn, ReadAbility, Inc

Photo Credits: © Chesapeake Images/Shutterstock Images, cover, 1, 17; © mharzl/Shutterstock Images, cover, 1, 19; © Catherine Murray/Shutterstock Images, cover, 1; © Iakov Filimonov/Shutterstock Images, 5; © Monkey Business Images/Shutterstock Images, 7; © joy_stockphoto/Shutterstock Images, 9; © Nikitina Olga/Shutterstock Images, 11; © Minerva Studio/Shutterstock Images, 13; © Sumikophoto/Shutterstock Images, 15; © shzumo/Shutterstock Images, 21

Library of Congress Cataloging-in-Publication Data

Masters, Nancy Robinson, author.
Flowers bloom / Nancy Robinson Masters.
pages cm.—(Tell me why)
Includes index.
ISBN 978-1-63362-610-2 (hardcover)—ISBN 978-1-63362-790-1 (pdf)—
ISBN 978-1-63362-700-0 (pbk.)—ISBN 978-1-63362-880-9 (ebook)
1. Flowers—Juvenile literature. I. Title. II. Series: Tell me why? (Cherry Lake Publishing)

QK653.M37 2015
581—dc23

2014049834

Cherry Lake Publishing would like to acknowledge the work of the Partnership for 21st Century Skills. Please visit *www.p21.org* for more information.

Printed in the United States of America
Corporate Graphics

Table of Contents

Choosing Seeds

"Are we going to the mall today?" Callie asked. She and her sister, Keesha, loved to go shopping with their grandmother.

"Yes, but first we are going to the seed store," Mrs. Lopez said. "You may each choose a small packet of flower seeds to buy."

At the store, there were hundreds of seed packets to choose from! Callie chose pansy seeds. Keesha chose daisy seeds.

LOOK!

Look at pictures in plant catalogs. Find flowers that bloom where you live.

Pictures on seed packages show the kinds of blooms the plant will make.

“Do you still want to go to the mall?” Mrs. Lopez asked after they checked out.

“Let’s head home and plant our seeds in the garden,” Callie said. “We want to see whose flowers bloom first.”

People of all ages enjoy watching different kinds of flowers grow and bloom.

From Buds To Blooms

All living things are made of **cells**. **Genes** are parts of cells. They are the codes in cells that control different things. For example, human genes control things like the color of our hair and the size of our feet.

Plant genes control how roots grow. They control the size of the stem and the shape of the leaves. They control where the **flower buds** form. Plant genes control how the blooms will look and smell.

Seeds grow into plants before they produce blooms.

Flowers have to wait until the air **temperature** is warm enough for them to bloom. But the blooming doesn't begin after just one warm day–it takes many warm days to encourage flowers to grow. The number of warm days needed, and how warm they must be, are different depending on the kind of flower.

Some flowers bloom once a day. Others bloom once a year or only at night. All flowers bloom in order to make new seeds.

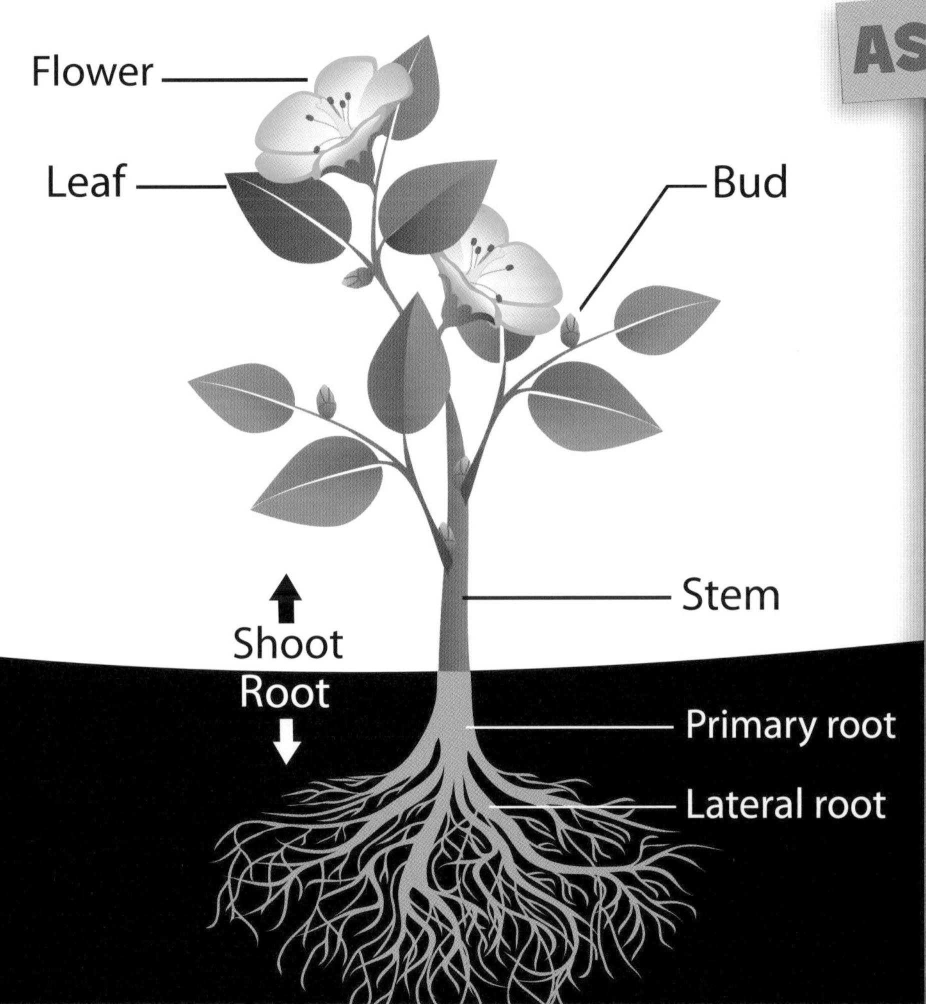

Plants need warm air temperatures in order to produce flower blooms.

ASK QUESTIONS!

Ask an adult to help you find a flower to study. Which parts from this diagram can you find? Which parts are still underground?

Plants make **pollen** for this purpose. Pollen is powdery material that **pollinators** carry from plant to plant. Wind, insects, birds, and animals are natural pollinators. Some pollinators, like bees and butterflies, carry pollen from flower to flower during the day. Bats and moths pollinate flowers that bloom at night. Flowers that are not pollinated do not make new seeds.

MAKE A GUESS!

Why do some people sneeze a lot when flowers bloom?

Warm temperatures in a greenhouse help flowers bloom.

Attracting Pollinators

Callie and Keesha made their choices by looking at the flower pictures on the seed sacks. How do pollinators choose which plants to pollinate?

Some flowers **attract** pollinators with the nectar they produce. Nectar is the sugary fluid made by flowers that birds and insects like to drink. Certain shapes of flowers attract some pollinators. Other blooms give off sweet odors. Plants can even trap pollinators inside their blooms!

Insects like honey bees pollinate flowers.

All flowers need soil, sunshine, and water to grow and bloom. It just takes one seed to make a plant. But a plant can make many flowers.

Hummingbirds are attracted to flowers that have nectar for them to drink.

When Will They Bloom?

Callie's pansies are **annuals**. The plants will bloom and make seeds for one year. Then Callie will need to plant more seeds to grow new plants.

Keesha's daisies are **perennials**. The same plants will bloom year after year.

Pansies are annuals. Each annual plant grows, blooms, and makes seeds for one year.

The girls looked at the labels on the seed sacks. The pansy seeds will become plants with blooms in about three months. It will take two months for daisy seeds to become plants with blooms. Keesha's daisies will bloom first!

Daisies are perennials. Each perennial plant can bloom year after year.

Think About It!

A plant nursery grows flowers for sale. Go online with an adult to see if there is a nursery in your neighborhood you could visit.

Would you choose annual or perennial flower seeds to plant? List three reasons for your choice.

What is the best thing to do if you see bees or wasps in a garden? Should you leave them alone? Why or why not?

Glossary

annuals (AN-yoo-uhlz) plants that only grow, bloom, and make seeds for one year

attract (uh-TRAKT) to cause something to move closer or touch

cells (SELZ) tiny units that are the building blocks of all living things

flower buds (FLOU-ur BUHDZ) blooms on plant stems getting ready to open

genes (JEENZ) tiny pieces of biological code in human and plant cells

perennials (puh-REN-ee-uhlz) plants that bloom year after year

pollen (PAH-luhn) powdery material from the male part of the flower, which when combined with the female part makes new seeds

pollinators (PAH-luh-nate-urz) wind, insects, birds, and animals that carry pollen from one flower to another

temperature (TEM-pur-uh-chur) the amount of heat or cold present

Find Out More

Books:

Colby, Jennifer. *Flowers*. Ann Arbor, MI: Cherry Lake Publishing, 2015.

Kalmon, Bobbie. *What Is Pollination?* New York: Crabtree Publishing, 2010.

Wade, Mary Dodson. *Flowers Bloom!* New York: Enslow Publishing, 2009.

Web Sites:

USDA—Forest Service: Just for Kids
www.fs.fed.us/wildflowers/kids/
Learn more about wildflowers through fun activities.

My First Garden: Planning My Garden
http://urbanext.illinois.edu/firstgarden/planning/
Read about how to plan and grow a flower garden (in English and Spanish).

Index

About the Author

Nancy Robinson Masters is the author of more than 40 children's books. She and her husband live on a farm near Abilene, Texas. They are airplane pilots who like to look for blooming flowers when they fly. Find out more about Nancy at www.nancyrobinsonmasters.com.